虎頭凳
Tiger Head Fighting Bench

Ancient Weapons of Kung Fu

虎頭凳
Tiger Head Fighting Bench

Ancient Weapons of Kung Fu

Paul Koh

Kung Fu in a Minute Publishing

KUNG FU IN A MINUTE

KUNG FU IN A MINUTE PUBLICATIONS

Copyright © 2025 Paul Koh

All rights reserved

DEDICATION

Dedicated to my Sifu

Tak Wah Eng

I am always in awe of his skill, knowledge, wisdom and generosity.

Dr. Mark Cheng

Foreword

There are few people in the world of Chinese martial arts whom I respect as much as Sifu Paul Koh.

When we first crossed paths in the mid-90s at a Chinese martial arts tournament in Ohio, I walked into a gymnasium and heard someone calling out commands and the names of techniques in beautiful Cantonese. As an American-born Chinese myself, I only had a tenuous grasp on Mandarin and negligible ability in Cantonese at the time. You can't even imagine my shock when I looked over to see a white man speaking those commands to a predominately African-American student body performing EXQUISITE technique with strong stances and crisp, fluid movements. His skill as an instructor was matched by his class, and recognizing me from the event program, he had his students bow and address me as "Cheng Sifu" before resuming their pre-competition warmup.

Fast-forward three decades, and that man, Sifu Paul Koh, has become like a big brother to me in the world of traditional Chinese martial arts. In every way that I've been privy to, Koh Sifu is a perfect example of the ideal in modern Kung Fu. Of Greek heritage himself, he's not only comfortable in the language of his ancestry and American homeland, but he's also an absolute joy to listen to when he breaks out into Cantonese. Similarly, watching him move and teach, Koh Sifu is neither one to force his students to fumble along blindly or to micro-manage them.

Recognizing that the new world of today's Information Age is one where the old ways of hyper-secrecy were a fast-track to the extinction of martial culture, Koh Sifu began working diligently to catalogue and communicate the Fu Jow Pai Kung Fu system of his master, the iconic Tak Wah Eng, and his grandmaster, the legendary Wai Hong. Koh Sifu's work has ensured that the Kung Fu of his lineage is protected, preserved, and promoted while so many other traditional schools have faded towards obscurity and their styles and systems have waned in popularity.

As traditional Chinese martial weaponry is a combination of classical battlefield armaments and "weapons of opportunity," any discussion of Kung Fu fighting should include the ability to take ordinary, everyday objects and employ them strategically and successfully as force multipliers in combat. Two such weapons which are archetypical in southern Chinese martial arts are the bench and butterfly knives.

Not to be confused with the Filipino balisong "butterfly knife," the Chinese butterfly knives are more like shorter broadswords, making them easier to carry and conceal. Some traditions even begin their forms with a nod to when the slimmer versions were carried as boot knives. Their robust build, just past the elbow length when held in inverted position, lends them towards more aggressive techniques and strategies than that of a thin stiletto type blade or a short dagger.

The Chinese bench resembles a common saw horse in construction, but with a lower profile allowing for ease in use as an easy to make, easy to move seat. For common folk, these were ubiquitous as seats in restaurants, tea houses, or any place where people gathered. While one might think of using such an implement as a desperate defensive tool or crude bludgeon, a skilled practitioner could turn the bench into a multipurpose tool for hooking, striking, sweeping, disarming, and close-quarter trapping.

For the truly skilled Kung Fu practitioner, the ability to adapt to any situation and apply the principles of one's chosen training system marked not only an excellent level of skill, but often the measure of life or death in more dangerous situations. The movements and techniques illustrated in this groundbreaking text not only archive the concepts of Fu Jow Pai, but they also serve as goalposts for us to strive towards with coordination using different implements. To see the skill sets around these classical weapons of opportunity brought to the general public by someone of Koh Sifu's caliber brings me not only great excitement as a lifelong student of the martial arts but also immeasurable joy as a Chinese-American.

Mark Cheng, L.Ac., Ph.D.
鄭馨民 醫師
@DrMarkCheng

- Former Contributing Editor: *Black Belt Magazine*
- Instructor: Inosanto Academy of Martial Arts
- Instructor: American Combat Shuai-Chiao Association
- Master SFG Instructor: Pavel Tsatsouline's StrongFirst
- Co-Founder: K3 Combat Movement Systems

Master Tak Wah Eng and Sifu Koh practicing bench vs. butterfly knives, late 1990s

INTRODUCTION

Over the years, I've had the opportunity to learn many bench sets from my teacher but in this one he has brought back the original or old-school flavor of the real Southern Cantonese old village style fighting bench. There are no flowery, unusable movements. Everything is geared towards the application of your Kung Fu technique. As my teacher said, "This is the real deal." This bench form is honoring the ancestors of our Kung Fu system and bringing back the old masters' idea and flavor. You can't get closer to authentic and original training from old masters than the bench because they weren't allowed to have access to other weapons, so they had to make due with what they had. All different household items became transformed into formidable weapons in their skillful hands. Fast forward three or four centuries and we are still able to apply this concept of taking the Kung Fu taught to you by your teacher and applying it to whatever you put your hands on. In this instance, it's the horse bench.

Many Southern systems utilize the horse bench as a fighting implement, but this set is ultmately unique because it's infused with the concept and principles of tiger claw movement and techniques as I have learned them. In this way, you can see the nature of the tiger hidden within the fighting bench. One becomes synonymous with the other. That's what makes this form unique, special and different. For the first time, the master is revealing his unique blend of tiger claw and fighting bench together. Its flavor and signature are authentic, more simple, more intense, direct and powerful, bringing it back home again to the original source of the tiger.

虎頭凳
Tiger Head Fighting Bench

The Kung Fu fighting bench is a well-known but not widely practiced weapon of the Chinese martial arts. This weapon is fashioned from the minds of village masters of Kung Fu predominantly in the regions of Guangzhou and other southern provinces. The southern systems of Kung Fu were never practiced by rich individuals of the Imperial Court, nor were the ancient masters granted access to military supplies or weapons of the day. This seeming negative was turned into a positive by the ingenious adaptation of many items at hand on a daily basis for the Southern practioners to defend themselves at a moment's notice. The pressures of daily life and survival forced these masters to take things as they came. Fighting could occur at any point in time, so one had to be able to make anything in their hands become a weapon to protect their lives. This is the genesis of this type of weaponry—farming tools, cups, bowls, chopsticks, cleavers, sharpened coins and, of course, the horse bench. Ingenuity, necessity and inspiration are all woven into the origins of the Kung Fu fighting bench.

The bench, as with many weapons found in the arsenal of Kung Fu, is highly specialized. It is neither a true weapon of military grade, nor is it merely furniture. Its commonplace location and ease of adaptability of many techniques has made it a favorite tool of many a village master. These teachers and fighters were able to employ and distill the techniques of other weapons and hand techniques into the use and practice of the fighting bench. Because it's a tool rather than a classic weapon of military grade, there's no standardization of the bench in terms of size or shape. Benches came in a variety of sizes, from a large bench used for sitting, working, sleeping and a smaller version used for cooking and handiwork. All can be used for fighting. Anything that these peasant masters could put thier hands on, they turned into a weapon. This is how highly skilled and innovative they were regardless of their lower social status. They were not rich people, but real people.

More than almost any other weapon, be it tool or weapon proper, the horse bench or Kung Fu bench uses its entire surface area for combat. The bench itself seems harmless enough, yet it the hands of a trained practitioner it becomes truly formidable. All aspects the bench, including the seat, corners, edges and legs can be used to engage with an armed opponent or multiple opponents. The bench utilizes its construction to defend and attack simultaneously with great power and strength. Actions of the fighting bench including striking in all directions—up, down, sideways, diagonal—sweeping, tripping, ensnaring, capturing, locking, shielding, blocking, pushing and barring.

The Tiger Head Fighting Bench is diverse in its actions and can encompass many different movements to be used in combat, including:

Battering
Barring
Hooking
Ensnaring
Swinging
Butting
Crashing
Shielding

Body integration is key to understanding how to bring out the hidden weapon in the tool. The tool by itself is only an inanimate object that has no mind or intention but rather something latent that it could potentially become. The integration of the understanding and viewpoint of the practioner's mind make the item change from an object with no purpose to a true weapon of Kung Fu. This is at the core of our philosophy, the principle that anything the Tiger Claw practitioner touches can and ultimately should be transformed into a formidable weapon. The source of this lays within the basic structure of the physical techniques in the system, and in the innovative mindset of the student. The individual should be able to think on his feet so to speak and improvise fighting strategies and techniques at will. Fine tuning of the body and mind are required to bring out the possibilities hidden within the Tiger Head Fighting Bench as with all weapons or non-weapons. Pens, pencil, keys, table, tray, chair, books and all other things can virtually be remade into weapons of Kung Fu within the Dynamic Art of the Tiger.

The reader should note when viewing these photos, they should read the sequence of movements from left to write and top of the page to bottom. The instructions denote the direction that each technique will face—front, back, left, right and the corresponding corner points.

Carry bench in crook of left arm, grasping leg of bench.

Draw fist to waist while executing tiger jump forward.

Land in horse with claw hand at right side, striking forward with bench on left.

Strike forward with right claw, drawing bench back in bow stance.

Step back with left foot into reverse cross stance, striking with right elbow to back.

Step out with left foot in side stance, raking diagonally with right claw facing front.

Execute high front kick and claw strike.

Step into reverse cross stance, executing claw strike.

Step to back with left foot, executing low bench block.

Follow up with right diagonal punch downward in bow stance.

Toss bench up with left hand.

Capture opposing side with right hand in single leg stance.

Step down and turn to back right corner with low thrusting strike with end of bench.

Step up into bow stance, striking upward to front left corner.

Step backwards, simultaneously striking with left and right corners of bench to finish in low striking position facing front.

Draw bench back and execute high right swing kick to front left corner.

Step down into left bow, executing upward strike with bench.

Step into reverse cross, turning body to execute high diagonal strike with bench to back.

Step forward into right horse, crashing downward with edge of bench.

Skip forward.

Land in right horse, thrusting bench forward.

Turn to face front.

Twist into left cross stance, drawing bench downward.

Step forward, wrapping seat of bench around in counterclockwise direction.

Strike directly upward with bench.

Shift to right in reverse bow, holding bench out perpendicular to floor in blocking position.

Strike down with right.

Immediately follow up with left.

Step forward into right tiger stance, striking upward with legs.

Execute high right front kick.

Step into right cross stance with horizontal battering strike.

Spin to face right side, bench in vertical blocking position.

Shift into right side stance, executing diagonal cutting strike with edge of bench.

Immediately shift to left side stance, executing diagonal cutting strike.

Step forward into left tiger stance, executing rising strike with seat of bench.

Execute high left front kick.

Step backward into right cross stance to face left side with thrusting bench strike.

Turn around, holding bench to left side in barring position.

While stepping forward, execute three consecutive upward slashing strikes with bench.

Spin around counterclockwise to face front.

Step into right bow with high bench block.

Retract bench to righthand side, stepping into left bow, executing downward smash with seat.

Drop down to floor, thrusting tiger tail kick to back.

Rise up from floor, turning to face back while hooking with legs.

Shift into reverse tiger stance, thrusting with high vertical strike with end of bench.

Turn to look forward, thrusting left tiger tail kick to front.

Step down into left bow, crashing directly down with edge of bench.

Flip bench while stepping into right cross stance facing front right corner, shielding body.

Invert bench again while stepping with left foot into cross stance.

Step out into right bow with crashing strike of bench seat.

Draw bench inward towards body while stepping into left cross stance facing front left corner.

Step around to face back right corner with right cross stance, flipping bench.

Step out into left bow, executing crashing strike with bench seat.

Facing the back, strike directly down with edge of bench.

Step forward executing double downward strikes, left hand, right hand.

25

Withdraw bench backwards and spin 180 degrees to face front.

Sink down into left kneeling stance with high bench block.

Drop left knee and execute stamping strike to ground with end of bench.

Pick up left leg to prepare for jump.

Jump over to face back.

Drop right knee and execute stamping strike to ground with end of bench.

Execute vertical thrusting strike in right kneeling stance directly back.

Turn body to left side, dropping down in low back stance, slamming bench on ground.

Raise bench immediately to left side, executing strike with legs.

Draw bench back to lefthand side, facing back.

Step over with right and left foot in zig-zag formation, ending in elongated right cross stance with rising bench strike.

Step over with left and right foot in zig-zag formation, ending in elongated left cross stance with rising bench strike.

Step back with right leg into left single-leg stance, drawing bench back and facing front.

Put left foot down and jump forward, moving bench in counterclockwise rotation.

Culminate jump into low back stance, bringing bench to ground.

Rise up, stepping forward into left tiger stance, executing uppercut strike with bench.

Step forward, executing hooking action with legs of bench and right leg.

Place right leg down, immediately executing spinning tiger tail kick.

Step down into half-horse, thrusting bench forward with both hands at mid-body level.

Shift into right reverse bow, drawing bench vertically to side of body.

Flip bench overhead onto left shoulder, stepping into right cross with right tiger claw to back.

Step up into half-horse, thrusting right tiger claw forward in closing salutation.

虎凳對雙刀
Bench vs Butterfly Knives

Matching sets are a stalwart feature in all Chinese martial art systems. Matching sets will range from short to long, simple as well as complex and will run the gambit between empty-hand against empty-hand, empty-hand against weapon and weapon versus weapon. The matching sets are used as a vehicle to bridge the gap between forms, either empty hand or weapon, and actual fighting sequences. This is extremely important when we speak about weaponry fighting. Fighting with weapons is even more succinct and deadly than empty hand fighting and this should be clearly evident to anyone that has wielded a weapon in their hand. In truth, when fighting with weapons, the altercation will be over in an instant. This being understood, the weapon matching sets are arranged in such a fashion that it will allow both sides to fully investigate and learn the possibilities of either weapon. Neither weapon is superior to the other but rather equal, all depending on the practitioner's skill and experience. This is what we seek to build up when we practice matching sets, the skill and experience of the individual.

Students' skills can be honed while practicing these preset choreographed routines within a certain margin of safety and allow the individuals to explore and understand the characteristics and nature of their weapons better without causing deadly harm to one another. This being understood, it by no means should be taken that the weapon matching set is just a playful routine. As stated earlier, when training with weapons, the slightest departure from the routine can cause irreparable damage to either player and this is what will be understood when we divide the weapon matching set into smaller segments and practice applications with them still under a controlled environment.

The specific matching set that we are looking at is two of the classical weapons found in the Southern Chinese martial art systems. The Southern Chinese martial arts have a wide variety of weapons and make full use of odd or esoteric weapons that may not necessarily be found on the battlefield. These two weapons, the bench and butterfly knives fall into this category, that of something more commonplace and everyday in usage. The butterfly knives are obviously an adaptation of a common man's tool, in this case the butcher's cleaver, outfitted with a special handguard to allow for more protection for the user's hand, as well as the ability to fold and open the knives in a reversed position, giving these twin blades a capacity to be used in short close-quarter combat, as well as longer range and allowing the practitioner to chop, cut, stab, ensnare and strike, as well as defend in both positions. The versatility of being able to fully extend and retract back to the forearm gives the butterfly knife practitioner a unique aspect in range, motion and techniques.

The fighting bench, on the contrary, is even more so an everyday household item that would be found in ancient times, as well as today. The bench is used not only as a surface to sit upon but also a workbench, a workhorse of sorts. At times, it has been used for sleeping on, as well as carrying items to and fro. The versatility of the bench in everyday usage is innumerable and is of course applied in the Southern tradition of martial arts. The bench is found in households, restaurants, tea houses and the like and would be a handy item that could be picked up by the Kung Fu expert and used immediately to defend oneself. The structure of the bench is solid and strong, allowing all the surfaces to be put into play when engaged in fighting. The multiple surfaces of the bench bring in a great array of ways to be applied. Firstly, the legs of the bench are conveniently held with right and left hand to block or engage with the opponent, be they with weapon or empty hand. This way of gripping the bench with the legs close to the base of the seat clearly can emulate the two hands that are used in empty hand form and quickly convert those movements into techniques with the bench. The legs are also uniquely used to ensnare, trap and hook, as well as strike with against an unarmed or armed opponent. This is put into full play in the tiger head fighting bench series. The fighting bench also will use the four corners of the seat to strike with, as well as the long-edged plank of the seat to incorporate offensive and defensive techniques, as well as the larger part of the seat platform itself. In this way, the tiger head fighting bench is an all around weapon that can be gripped close to the body, used for striking left, right, high, middle or low, as well as front and back or be fully extended to the longest possible reach, executing long swinging hooking movements that also emulate long hand or long range techniques of the empty hand movements found in our forms. In this way, the bench is convenient, easily converted and becomes an incredible fighting tool that is found in traditional Southern systems of Chinese martial arts.

The matching set presented here in this text is an application of all these amazing attributes that these two classic Southern Kung Fu weapons embody. As with all matching or fighting sets, they are initially set forth to teach the students the basic techniques and principles of each weapon, as well as how they may interact with one another. This is well displayed in this matching set, yet much much more is at play. The mixture of single weapon vs double, the comparison of the hard and the soft, the flexible and the stalwart, ranges of close and far, sharp edges against the dull—all this brings a out a greater understanding of not only the adaptability of the bench but also the skill of the double knives, culminating in a crescendo of empty hand tiger claw techniques from both combatants displaying the immense Kung Fu legacy of our martial ancestors.

This matching set is based on equal footing between the two weapons and shows the full range and capacity of these two weapons in their ability to fight in short- or long-range sequences, as well as jumping and rolling techniques that are applied with both weapons. The matching set equally displays both weapons at their fullest potential, showing a myriad of offensive and defensive techniques that can be utilized on both sides, as well as the capacity to disarm one another and continue the battle empty-handed to its final conclusion of both being well-matched for one another. The ancient techniques of Kung Fu masters of a gone by era are in full display in this matching set, and it is an all-encompassing compilation of techniques of the fighting bench and butterfly knives.

The reader will see each side of the matching set displayed as its own form, followed by the set itself with both partners interacting. For best results, individuals attempting to learn these two separate sets, as well as the matching piece, should strive to only attempt several movements at a time to better digest and understand the material. Over time, with consistent practice and patience, one should be able to do more and more.

Stamp forward with both feet together, executing 90 degree low punch.

Shift backward into right cross stance executing tiger claw strike to back.

Step forward into left tiger stance, presenting tiger claw salutation.

Slide left foot back and execute inside right crescent kick, spinning counterclockwise.

Finish rotation, ending in left side stance position with right tiger claw striking, facing right side.

Shift to right side stance, executing inside left chopping strike.

Turn to left bow stance, right tiger claw strike facing front.

Shift into low back stance, double tiger claw behind.

Grasp bench with right hand and kick up with right foot.

Grasp bench with both hands, assuming half-horse position, thrusting forward in horizontal strike.

Shift back to left side stance, holding bench in vertical position.

Shift back to right side stance, holding bench in vertical position.

Turn body clockwise, wrapping bench around head and torso.

Completing rotation, step back with left foot into right cross stance, drawing bench back.

Execute high left front kick.

Draw left leg back in preparation for cartwheel jump forward.

Finish spin forward, landing in low left back stance, slamming bench to ground.

Rise up, shifting into left side stance, extending bench out to left.

Move into left tiger stance, bringing bench to right side.

Step over with left foot into cross stance on a 45 degree angle, bringing bench down.

Step forward with right foot, drawing bench to right side, low.

Step out with left foot into tiger stance, holding bench with legs facing outward.

Shift up, executing high block on 45 degree angle.

Shift to right side, executing vertical block with left hand high.

Slide left foot back into horse stance, flipping bench to block center of body in vertical position, right hand up.

Move into right bow stance, executing low block downward, facing front.

Step back into half-horse position, drawing bench backward.

Advance forward into right bow, blocking up.

Raise right leg in single leg stance to begin counterclockwise turn.

Complete rotation, landing in left kneeling stance with bench shielding in vertical position.

Step over with right foot into cross stance, bringing bench downward.

Cross over with left foot, drawing bench to left side high.

Step out with right foot, sliding left back, bringing bench up over head.

Draw bench back to right side, preparing to turn and jump.

Spin towards left side, jumping backward, bench wrapping around head.

Complete spinning jump, landing in left kneeling stance, driving bench vertically down to ground.

Flip bench over to other side, stamping end onto ground.

Drop bench down to ground with left knee, executing right tiger tail kick.

Bring right leg down, assuming kneeling ready stance position.

Rise up and shift to left in left side stance.

From side stance position, step over with right foot, striking high with bench.

Raise left leg in single leg stance, blocking with bench legs.

Immediately step down to right cross stance, bench defending low.

Shift out to left side stance, raising bench to defend high.

Slide left foot back, throwing bench backward to attack.

Bring bench down. Step forward with left foot, raising right, executing spinning hook strike with bench legs.

Complete hooking rotation strike with bench. Raise left leg.

Bring left leg down into low left side stance, executing right downward strike with end of bench.

Draw bench back into right tiger stance.

With right hand, sweep outward with bench.

Follow through with sweeping rotation.

Drop down to right knee, bending backward with high block.

In left kneeling position, bring bench downward to righthand side.

Immediately, draw bench overhead, raising left leg.

Step down into left horse with downward crashing strike with edge.

Step to righthand side, sliding left foot back into cross stance, flipping bench.

Step to lefthand side, sliding right foot back into cross stance, flipping bench.

Execute right sweeping kick, rotating body around.

Complete rotation, landing in left side stance, vertical blocking position with bench.

Draw bench back, lifting left leg to jump forward.

Jump forward into left horse, executing thrusting strike with seat of bench.

Shift back slightly, drawing bench back to right side.

Shift to right side stance, flipping bench to right, blocking vertically.

Step forward, hooking with legs of bench.

Spin around, executing high tiger tail kick.

Immediately execute counterclockwise cartwheel turn.

Land on ground, driving bench downward.

Drop back onto both hands, executing right tiger tail kick from ground.

Place right foot down, executing double tiger claw strike forward.

Rise up into right side stance, rolling claw to right.

Shift to left side stance, rolling claw to left.

Raise right leg into single leg stance with double tiger claw attack.

Step down into left side tiger stance with high and low tiger claw deflection.

Shift back into right horse with high right tiger claw.

Draw right hand back. Shift to right side stance, thrusting out left tiger claw strike.

Shift back into low horse, executing low X-block.

Shift to right, executing left tearing tiger claw strike. Follow through with full rotation.

Complete rotation. Drop down to ground, executing right tiger tail kick.

Roll backwards.

Turn around in right kneeling stance with double rising bridge hand block.

Shift right leg forward, twisting hands into cross tiger claw position.

Shift onto right knee, executing left 360 degree left sweep.

Rise up into right side stance in finger strike position.

Shift to left, executing diagonal downward swinging strike.

Retract hands backward, lifting left leg into single leg stance.

Step forward and down into low left back stance with double knife hand chop.

Turn clockwise, jumping backward.

Land in right bow stance with low X-block.

Execute triple high and low knife hand blocks.

Step forward with left foot into cross stance, executing diagonal slashing chop.

Step out to right side with tearing tiger claw strike.

Execute right and left rotating outside forearm block.

Follow through with double uppercut to right side.

Execute right outside crescent kick.

Place right foot down. Spin to right, turning and executing inside tiger claw strike.

Step back into left cross stance with double claw back.

Step up with kick right front kick, slapping instep with left palm.

Step backward into right cross stance, whipping out double phoenix eye strike.

Step forward into half-horse with double eagle claw strike.

Shift to left with outer forearm block.

Step forward with right tiger claw to groin.

Execute right outside crescent kick.

Shift back into left tiger stance, extending tiger claws high and low.

Sweep with right foot, simultaneously executing tiger claw tear, spinning around counterclockwise.

Complete rotation, landing in low right back stance, double knife hand position.

Rise up into left tiger stance with right and left claws extended.

Sink both bridge hands, extending fingers up and out.

Bring both hands to center of chest and close with tiger claw salutation.

Cradling double knives in left hand, stamp forward with both feet together, executing 90 degree low punch.

Shift backward into right cross stance executing tiger claw strike to back.

Step forward into left tiger stance, presenting tiger claw salutation.

Slide left foot back and execute inside right crescent kick, spinning counterclockwise.

Finish rotation, ending in left side stance position with right tiger claw striking, facing right side.

Shift to right side stance, executing inside left chopping strike.

Turn to left bow stance, right tiger claw strike facing front.

Shift into low back stance, double tiger claw behind.

Spin to right, opening up knives.

Finish rotation, crossing left knife over right.

Step out into left horse stance, striking with both knives forward.

Shift to left, slashing upward with both knives, parallel to floor.

Shift to right, slashing upward with both knives.

Turn to left, wrapping both knives around head.

Complete turn, stepping back into cross stance, knives in vertical position.

Face forward, executing high left front kick.

Retract left leg, preparing for cartwheel jump.

Execute cartwheel rotation ending in right cross stance with knives crossed in front.

Flip left knife into reverse position. Step forward into left tiger stance, knives touching at pommel.

Step over into right cross stance.

Step over into left cross stance.

Bring knives over to left side, drawing right leg up into single leg stance position.

Step down into right bow, executing high right chop.

Fold right knife into reverse position, shifting to right side stance with left cross-cutting strike.

Step forward into right horse with right cross-cut to center.

Open left knife, stepping forward into left bow with stabbing strike low.

Open right knife, slashing with both overhead while executing right inside sweeping kick.

Complete rotation, ending in low left horse with double upper slashing knives.

Step over with right foot into cross stance, executing double chop to front and back.

Turn to left, rotating knives overhead.

Complete rotation into left kneeling stance, both knives at left side.

Cross over with right foot, executing high and low cut to right.

Cross over with left foot, executing high and low cut to left.

Step forward into right cross stance, tucking right knife under left.

Drop down to floor, executing shoulder roll.

Complete roll into left kneeling position, executing right horizontal cut.

Turn body around to right kneeling position, executing left horizontal cut.

Drop down, executing backward shoulder roll.

Complete roll, landing in right kneeling stance, left knife down, right knife on shoulder

Step laterally forward and back, alternating knives up and down.

Shift to left with right slash.

Shift to right with left slash.

Rise up, stepping over into left cross stance with double low chop.

Rise up, stepping over into right cross stance with double low chop.

Fold both knives into reverse position.

Rise up into left single leg stance, knives blocking middle and high.

Rise up into left single leg stance, knives blocking middle and high.

Raise right leg to execute sweeping kick while slashing with right knife in elbow cut.

Complete rotation, stepping back into right cross stance, left knife on back, right knife at shoulder.

Take off from ground.

Jump over.

Spin around.

Open knives in single leg stance.

Step forward into left front stance with double chop down.

Shift back into half-horse.

Flower knives downward towards right. Spin around backwards.

Raise knives.

Bring knives down in double block.

Draw both knives outward to right and left sides.

Step back into right cross stance, flipping knives to right, parallel to each other.

Step into left cross stance, flipping knives to right, parallel to each other.

Bring knives back to right side, executing high left sweeping kick.

Spin to right side.

Land in right bow stance, executing double cut to back.

Fold knives into reverse position. Pick up right foot to turn around towards left.

Complete rotation into left tiger stance with high and mid block.

Open left knife, stepping into right single leg stance with high slash attack.

Step down into right cross with right horizontal chop.

Open both knives and spin towards right side.

Complete rotation, bringing right foot back into low back stance, extending both knives out.

Shift body to right kneeling stance, knives in high-low guard position.

Bring right foot back into left kneeling stance with high X-block.

Place knives down to left side. Execute right tiger tail kick from ground.

Put right foot down. Execute double tiger claw strike forward.

Rise up with high low tiger claw to left.

Shift to right, executing same.

Bring right foot forward into tiger stance with high tiger claw attack.

Throw high right roundhouse kick.

Step forward into right cross stance with right whip punch.

Shift right leg to side stance, bringing both hands into long bridge position.

Go down into left kneeling stance with high block and stabbing tiger fist strike.

Rise up into right side stance with left inside forearm deflection.

Shift to left bow stance with rising right tiger claw.

Step forward into left cross stance.

Execute high front kick, slapping instep with right hand.

Jump forward.

Land in right bow stance with double backfist strike.

Draw right leg up into single leg stance, crossing both hands in front of chest.

Shift back and display high and mid-level bridge hand position.

Shift to left. Draw hands in clockwise position, right back of hand clapping into left palm.

Execute right outside crescent kick.

Place right foot down. Spin to right, turning and executing inside tiger claw strike.

Step back into left cross stance with double claw back.

Step up with kick right front kick, slapping instep with left palm.

Step backward into right cross stance, whipping out double phoenix eye strike.

Step forward into half-horse with double eagle claw strike.

Shift to left with outer forearm block.

Execute right and left rotating outside forearm block.

Follow through with double uppercut to right side.

Shift to left, executing diagonal downward swinging strike.

Retract hands backward, lifting left leg into single leg stance.

Step forward and down into low left back stance with double knife hand chop.

Land in right bow stance with low X-block.

Execute triple high and low knife hand blocks.

Step forward with left foot into cross stance, executing diagonal slashing chop.

Step out to right side with tearing tiger claw strike.

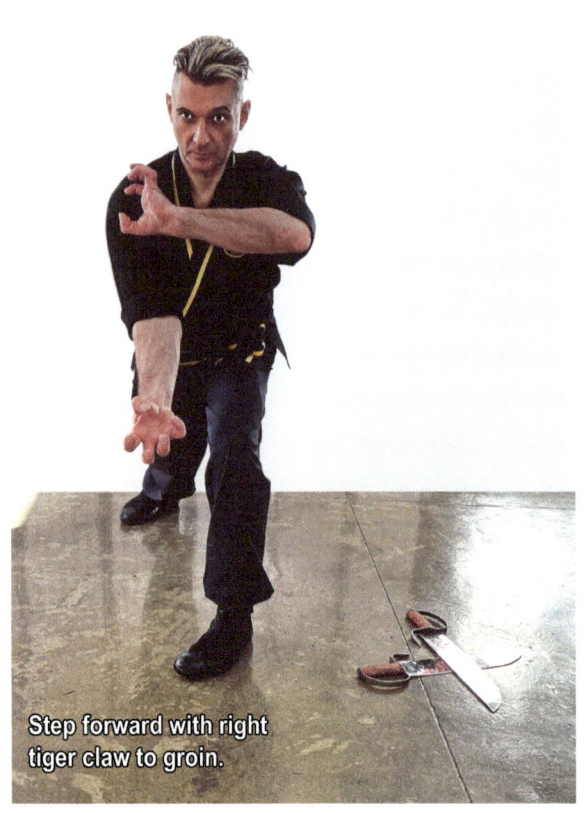
Step forward with right tiger claw to groin.

Execute right outside crescent kick.

Shift back into left tiger stance, extending tiger claws high and low.

Sweep with right foot, simultaneously executing tiger claw tear, spinning around counterclockwise.

Complete rotation, landing in low right back stance, double knife hand position.

Rise up into left tiger stance with right and left claws extended.

Sink both bridge hands, extending fingers up and out.

Bring both hands to center of chest and close with tiger claw salutation.

Both partners begin by stamping forward, feet together, executing low punch.

Both partners shift back into cross stance, executing tiger claw strike.

Both partners step forward into left tiger stance with tiger claw salutation.

Both partners turn, executing inside crescent kick.

Both partners land in left side stance, executing right tiger claw strike.

Both partners shift to right side stance with left chopping strike.

Both partners turn towards each other with right tiger claw strike.

Both partners shift into back stance with double claw strike behind.

BENCH rises up and kicks up bench with right foot, grasping with right and then left hand. KNIVES opens both knives, preparing to turn around.

Both partners assume ready position.

Both partners shift to left, brandishing weapons.

Both partners shift to right, brandishing weapons.

Both partners turn around, wrapping weapons around head.

Both partners complete turn, stepping back into cross stance.

Both partners execute left front kick.

Both partners assume left single leg stance.

Both partners execute circular jump towards one another.

Both partners land with weapons on the ground at the ready.

BENCH rises up from floor, shifting to left side, extending bench outward. **KNIVES** rises up from floor, stepping into left tiger stance, holding knives pommel to pommel.

BENCH shifts into left tiger stance, brandishing bench to right side. **KNIVES** steps back into right cross stance.

BENCH steps over with right foot into cross stance. **KNIVES** steps over with left foot into cross stance, encircling one another.

Both partners face one another to begin encounter.

Both partners engage. **KNIVES** steps forward with right, chopping high, knife fully extended, left in reverse position at waist. **BENCH** steps forward with left, blocking high.

KNIVES retracts right knife into folded position, attacking with left knife in horizontal slash. **BENCH** shifts body to block with bench in vertical position.

KNIVES shifts into horse stance, attacking with right horizontal slash. **BENCH** assumes horse stance, flipping bench, blocking in vertical position.

KNIVES steps forward into left bow stance, opening left knife, executing stabbing strike. **BENCH** slides left foot back, blocking downward with seat of bench.

KNIVES opens both knives, slashing high. **BENCH** ducks down, sweeping at **KNIVES**' right leg with legs of bench.

Both partners continue rotation, turning to face one another. **KNIVES** executes double under cut. **BENCH** executes down block.

KNIVES steps over with right leg into cross stance, chopping high. **BENCH** steps into right bow stance, executing high block.

Both partners spin away from one another to disengage and assume new ready position in kneeling stances.

Both partners criss-cross one another in zig-zag steps.

BENCH draws left leg backward into high cross position. **KNIVES** steps forward, drawing knives into body.

KNIVES rolls on floor, initiating attack. **BENCH** turns away, encircling head with bench, dropping down to bar chopping strike of **KNIVES**.

KNIVES flips body to strike with left knife. **BENCH** shifts, flipping bench to block chopping strike.

BENCH uses right hand side to knock knife down to floor, executing tiger tail kick. **KNIVES** retracts knife, ducks under kick and rolls away.

KNIVES completes rolling escape. Both partners assume new ready position. **KNIVES** executes up and down crouching step.

BENCH rises up, assuming left side stance.

KNIVES turns towards left side, slashing diagonally with right knife. **BENCH** evades, sliding left foot back, raising bench high.

KNIVES turns towards right side, slashing diagonally with left knife. **BENCH** evades, raising left leg into single leg stance, barring cut with bench.

KNIVES crosses over with left foot, executing double knife cut.
BENCH side steps with right cross step.

KNIVES crosses over with right foot, executing double knife cut.
BENCH steps out with left leg, bringing bench to left side.

KNIVES folds in both knives, encircling torso, defending high and low position in single leg stance.
BENCH crosses over with right foot, raising bench overhead in diagonal position.

Both partners prepare for simultaneous attack, spinning into each other.

BENCH completes horizontal rotation, striking downward at **KNIVES**. **KNIVES** empties out, stepping back into right cross stance to evade.

BENCH recoils into right tiger stance, executing low single-handed sweep with legs of bench. **KNIVES** jumps over sweeping attack, turning around to execute double high chop. **BENCH**, coming out of full rotating sweep, drops into left kneeling stance, arching back to block with seat of bench.

BENCH rolls knives over to righthand side by tilting bench.

KNIVES retracts and folds both knives inward, spinning away to avoid **BENCH**.

BENCH steps back, raising bench up to smash down on **KNIVES**.

BENCH crashes down with seat of bench. **KNIVES** comes over, suppressing strike downward with knives situated in center.

KNIVES extends both knives to either side of bench legs.

Both partners apply tension inward and outward with weapons, simultaneously rolling to right and left.

BENCH executes right low sweeping kick. **KNIVES** evades with left foot. Both partners complete spinning rotation.

KNIVES shifts into right horse stance, executing double horizontal chop. **BENCH** steps into left horse, blocking in vertical position.

BENCH attempts to brush off double knife attack with downward strike, recoiling into left single leg stance for next attack. **KNIVES** turns towards left, rotating away to avoid oncoming while folding knives into reverse position.

BENCH launches forward with thrusting left strike. **KNIVES** deflects with left forearm block.

KNIVES opens right knife, slashing at head of **BENCH**. **BENCH** ducks under.

KNIVES follows through, stepping over, executing inward right chop. **BENCH** shifts to right side, blocking with seat of bench, vertically.

BENCH ensnares knife with legs of bench, spinning around, executing tiger tail kick with left leg. **KNIVES** ducks under to avoid kick.

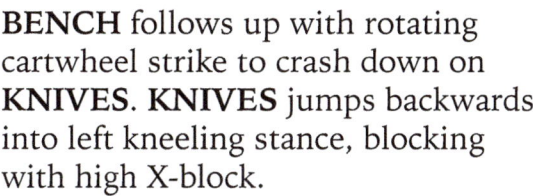

BENCH follows up with rotating cartwheel strike to crash down on KNIVES. KNIVES jumps backwards into left kneeling stance, blocking with high X-block.

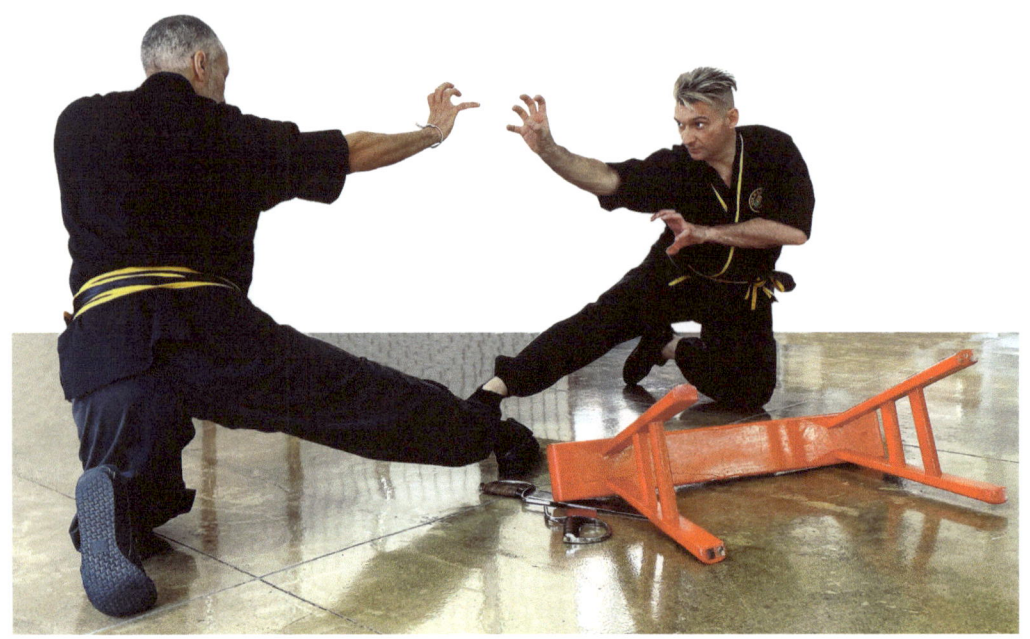

BENCH uses bench to drive knives downward to floor, disarming **KNIVES**. Both partners assume double tiger claw ready position in kneeling stance.

Both partners execute on the floor tiger tail kick.

Both partners rise up. **BENCH** executes double tiger claw in right side stance. **KNIVES** executes rotating tiger claw in half-horse.

BENCH executes double tiger claw in left side stance. **KNIVES** executes rotating tiger claw a second time.

Both partners execute high and low tiger claw attack.

Both partners spin counterclockwise. **KNIVES** executes right roundhouse kick. **BENCH** blocks with high low tiger claw.

KNIVES spins around clockwise, executing right whip punch. **BENCH** side steps towards right to capture strike.

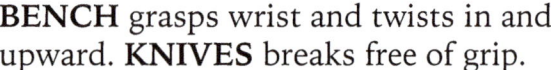

BENCH grasps wrist and twists in and upward. **KNIVES** breaks free of grip.

BENCH immediately executes left tiger claw strike. **KNIVES** shifts to side, blocking with long bridge hand.

KNIVES attacks with low stabbing strike. BENCH blocks with downward X-block.

BENCH immediately follows up with left slashing tiger claw strike to head of KNIVES. KNIVES ducks down to avoid.

BENCH simultaneously drops to floor, executing right tiger tail kick. KNIVES rises up, deflecting with left inner forearm black.

141

KNIVES swiftly scoops up kick with right rising forearm.

BENCH escapes with forward roll. **KNIVES** executes high right front kick.

As **BENCH** rolls away, **KNIVES** attempts to catch up, jumping forward.

KNIVES lands in right bow stance with double backfist strike. **BENCH** in right kneeling stance executes double high block.

BENCH grasps both strikes and attempts to ensnare **KNIVES**.

KNIVES pulls free. Both partners assume ready position.

BENCH immediately executes left floor sweep. **KNIVES** evades with right single leg stance.

Both partners continue rotation and face each other in new ready position.

BENCH executes diagonal downward swinging punch. **KNIVES** rotates hands clockwise into high left palm strike.

KNIVES executes right outside crescent kick. **BENCH** shifts backward into left single leg stance to evade.

BENCH moves forward into low right foot sweep with double knife hand strike. **KNIVES** turns body away, executing high tiger claw strike.

BENCH shifts forward into right bow stance, executing low X-block. **KNIVES** steps back into low right cross stance with double claw strike.

BENCH executes high and low knife hand block to both sides. **KNIVES** executes high right front kick.

BENCH executes third consecutive high low knife hand blocks. **KNIVES** steps into right cross stance, executing double phoenix eye strike.

BENCH steps forward with left foot, executing diagonal knife hand chop. **KNIVES** steps forward into right half-horse position with double claw strike.

BENCH side steps to right, executing tiger claw strike.
KNIVES shifts to left with outside forearm block.

Both partners shift to right then left with rotating forearm block.

Both partners shift to right, executing double uppercut.

KNIVES executes diagonal downward swinging punch. **BENCH** rotates hands clockwise into high left palm strike.

BENCH executes right outside crescent kick. **KNIVES** shifts backward into left single leg stance to evade.

KNIVES moves forward into low right foot sweep with double knife hand strike. **BENCH** steps down into right bow stance, executing high tiger claw strike.

KNIVES shifts forward into right bow stance, executing low X-block. **BENCH** steps back into low right cross stance with double claw strike.

KNIVES executes high and low knife hand block to both sides. **BENCH** executes high right front kick.

KNIVES executes third consecutive high low knife hand blocks. **BENCH** steps into right cross stance, executing double phoenix eye strike.

BENCH steps forward with left foot, executing diagonal knife hand chop. **KNIVES** steps forward into right half-horse position with double claw strike.

KNIVES side steps to right, executing tiger claw strike. **BENCH** shifts to left with outside forearm block.

Both partners step towards one another, executing low tiger claw strike to groin.

Both partners execute right outside crescent kick.

Both partners shift back into left toe stance with tiger claws high and low.

Both partners execute inside hook kick with tiger claw swipe, spinning counterclockwise.

Both partners shift into back stance.

Both partners rise up into left tiger stance with downward facing claw on either side.

Both partners sink right and left bridge hands.

Both partners culminate matching set in tiger claw salutation.

兵器散手
Weapon Applications

    Ingenuity is the one word to sum up the tiger fighting bench and its techniques. The ability to use this everyday item as a formidable weapon is beyond amazing. The mind does not quickly see common items as formidable enough to deal with other proper weapons used in times past nor today. Imagine looking at the common stool as a true weapon, yet it is in the hands of the Kung Fu expert. With all fighting techniques and situations we must bear in mind that nothing is foolproof nor guaranteed. The Kung Fu exponent must have a firm grasp of his basic understanding of how to fight with a variety of weapons and to fight in general. Therefore, one can take nothing for granted and must be ever vigilant on the situation and circumstances. With this firmly understood, we can look at the advantages and disadvantages of the fighting bench. Knowing both ends of the spectrum, we will be better equipped to apply the unique and simple tool as a formidable weapon.

    The bench itself provides a great deal of defensive and offensive capabilities as its size and length can keep the armed opponent at a greater distance, while at the same time act as a makeshift shield to hide behind to fend off blows from incoming attacks. The seat of the bench can easily be seen as shielding the body of the practitioner by hiding behind it, as well as crouching down to cover the practitioner's entire body similar to a rattan shield. This attribute of the bench is one of the important aspects of this weapon that will allow us to further explore the more aggressive attacking techniques available to the bench.

    The entire surface area of the bench can and will be used to fight from the seat, edges, legs and so on. The length and breadth of seat and edges of the bench will be employed for for attacking and striking. Striking with the bench can be seen akin to the double-head staff in many of its techniques, but also follows the movement of a battering ram of sorts that can crash and smash down on the opponent, creating great damage. The legs, as unassuming as they may appear, are powerful offensive weapons. A mere turn of the body will cause the legs to strike out and hit anything that is within range. This in of itself presents the practitioner with four additional arms with no extra effort. The unassuming nature of the bench is the secret to the individual's application of said attacking techniques. Above and beyond the ability to strike with these new limbs, the bench can also capture, trap, entangle, disarm or lock the opponents weapon, as well.

    Further understanding is that the bench is an extension of the body of the tiger claw practitioner, and he is able to incorporate empty hand and foot techniques with the bench during combat. Punches, kicks and the like are easily combined with bench techniques to complement our arsenal of movement.

    The practitioner utilizing the bench must be aware at all times of the position of his hands. These are the most vulnerable to attack from any opponent regardless of weaponry. The hands holding the bench are slightly exposed to cuts and strikes from other weapons. Therefore, consideration must be given to how, when, where, and why techniques are executed to protect oneself. The bench can be utilized with both hands simultaneously for a secure grip for defensive and offensive maneuvers or utilized single-handedly for swinging techniques or hooking movements to disarm or strike out at singular or multiple opponents. Make no mistake, the bench can be used against several unarmed individuals or opponents with weapons. This versatility makes the bench almost indispensable in one's repertoire of makeshift weapons.

The techniques displayed in this chapter are set forth in a freeform mode with little or no preparation whatsoever. No clear laid out plan was made as to how the bench would be utilized but rather left up to spontaneous action. Pictured here are the bench against the staff as a long weapon, saber and three-sectional staff.

The application techniques are broken up from varying series of movements taken from the form directly and applied at will, hopefully displaying and covering the gambit of techniques that the bench can apply but by no means showing all its versatility. The techniques in this chapter conform closely to many of the tiger claw principles, as well as physical movements themselves and therefore are a true extension of the Dynamic Art of the Tiger.

凳對單刀
Bench vs. Saber—Application #1

Both partners square off.

SABER attacks with thrust.

BENCH deflects downward with left side.

SABER steps forward with left, thrusting again. BENCH deflects inward with right side, stepping forward with right leg to attack with right end of bench. SABER shifts to right side to block bench with flat of saber.

BENCH follows up with spinning attack overhead. SABER ducks down to evade.

SABER attacks with thrust. **BENCH** ensnares with legs of bench.

BENCH immediately spins to right, using legs to trip and sweep **SABER**'s right leg, taking him to the ground.

Uprooting **SABER**, **BENCH** immediately jumps forward with thrusting strike to finish opponent.

凳對單刀
BENCH VS. SABER—APPLICATION #2

Both partners square off.

SABER attacks with overhead chop. **BENCH** blocks high.

SABER steps into cross stance with low chop. **BENCH** deflects down.

SABER spins around with high downward chop.

BENCH intercepts chop with center of seat, steps forward, bringing bench over and crashing down with right side strike to head of attacker.

凳對單刀
Bench vs. Saber—Application #3

Both partners square off.

SABER steps forward with chop to center. **BENCH** steps with cross stance to side, blocking attack.

SABER follows up with second lateral cut. **BENCH** shifts into horse stance, blocking with center of bench.

SABER advances with overhead downward chop. **BENCH** shifts, retracting bench to field attack.

BENCH ensnares saber and hand with legs of bench, forcing weapon and opponent down to ground.

BENCH immediately follows up with right tiger tail kick to face.

凳對棍
Bench vs. Staff—Application #1

Both partners square off.

STAFF attacks with right side strike. **BENCH** shifts to left, blocking with seat.

STAFF shifts back, striking with left. **BENCH** flips and turns into cross stance, blocking strike.

STAFF strikes overhead with right. **BENCH** intercepts, taking staff down.

BENCH follows up with right cross strike to head.

BENCH ensnares and strikes opponent's head with legs and seat of bench.

BENCH drives down opponent to ground, following up with finishing tiger claw strike.

凳對棍
Bench vs. Staff—Application #2

Both partners square off.

STAFF steps forward with folding overhand strike. **BENCH** intercepts with seat and drives down towards right side.

STAFF spins to left with low diagonal strike. **BENCH** steps back into cross stance, deflecting strike.

STAFF immediately spins to other side.

STAFF completes turn, striking low.
BENCH shifts into cross stance to parry.

STAFF retracts weapon in preparation for attack.

STAFF thrusts out stabbing strike with weapon. **BENCH** side steps, moving forward, ensnaring staff with legs of bench.

BENCH, while controlling staff, immediately spins body, executing tiger tail kick to head of opponent.

凳對棍
Bench vs. Staff—Application #3

Both partners square off.

STAFF executes thrust strike with weapon. **BENCH** deflects with outer section of legs.

STAFF recoils and executes second thrust. **BENCH** again uses outer surface of legs to block stabbing strike.

STAFF attempts third stabbing strike. **BENCH** shifts back, bringing staff downward to capture weapon.

BENCH disarms opponent with legs, striking staff and hands.

After disarming **STAFF**, **BENCH** follows up with finishing strike.

凳對三節棍
Bench vs. 3-Section Staff—Application #1

Both partners square off.

BENCH launches forward with thrusting strike. **3-SECTION** shifts backwards, block with left stick.

3-SECTION steps forward, striking overhead with right stick. **BENCH** shifts backwards, deflecting high with seat of bench.

3-SECTION steps over into cross stance with low cutting strike with right stick. **BENCH** steps forward into cross stance, sweeping away strike with legs of bench.

3-SECTION turns to his left, executing overhand strike with right stick. **BENCH** turns around, drawing left foot down, dropping to ground to evade.

3-SECTION steps out with right foot, executing low left strike with stick. **BENCH** rises up, deflecting with end of bench.

BENCH follows up directly, moving into cross stance with thrusting strike to head of opponent.

凳對三節棍
Bench vs. 3-Section Staff—Application #2

Both partners square off.

3-SECTION steps forward with overhead whipping strike with right stick. **BENCH** ducks to evade.

3-SECTION follows up immediately with low whipping strike to legs. **BENCH** jumps up to escape.

3-SECTION retracts weapon and executes mid-level whipping strike. **BENCH** moves to side, using to seat to block.

BENCH immediately spins towards opponent with overhead horizontal strike.

3-SECTION ducks down to evade horizontal strike.

BENCH completes turn, drawing bench back to right side.

BENCH immediately follows through with left thrusting strike to head, retracting bench and executes second downward strike to knee joint of opponent to finish encounter.

凳對三節棍
Bench vs. 3-Section Staff—Application #3

Both partners square off.

BENCH executes rising upward strike with left. **3-SECTION** deflects with right stick.

BENCH follows up with right attack. **3-SECTION** deflects with left.

BENCH executes third rising upward strike. **3-SECTION** shifts to side, deflecting with right stick.

BENCH deflects 3-section staff downward.

BENCH steps in, executing right sweeping kick to leg of opponent. **3-SECTION** evades, raising right leg.

BENCH completes spin on sweeping kick, turning to face opponent.
3-SECTION jumps forward to attack.

3-SECTION strikes with both right and left sticks overhead. BENCH executes high block with seat.

BENCH rolls off double stick attack to right side with seat of bench, driving 3-SECTION down to ground.

Before opponent can recover, **BENCH** executes downward crashing blow to head and neck.

月圓虎嘯圖 *Tiger Roars with the Rise of the Moon* by George Wong

CONCLUSION

The ancient weapons of the art of Kung Fu are many and diverse, yet there are many hidden or not-quite weapons that have been improvised as we have seen in this text. The bench is one such implement that has been improvised in the most ingenious manner. Initially, one looks upon this household item of a bygone era and will not think much of it. But when truly spending the time and effort, which is the defining ingredient of Kung Fu, one will discover a very amazing weapon indeed. The bench allows the practitioner's interpretation of a wide variety of techniques from a myriad of weapons and empty hand techniques, especially from the standpoint of a true practioner of the Tiger Claw Kung Fu System. With great practice, the bench can be transformed into a formidable weapon that provides great defensive and offensive components intertwined as one. Incorporated into this marvelous tool can be seen elements of the Dynamic Art of the Tiger—striking, blocking, locking, disarming. The entire surface of the bench is the weapon, from the seat to the ends and edges to the legs. No surface is without use and application.

Weapons aren't always what they seem to be. That is to say, they may not be a recognized classical weapon from the battlefield. Many weapons in Kung Fu are improvised or concocted on the spot either by sheer accident or luck. Speaking of luck, there truly isn't any. What seems lucky to someone looking in is truly the deep understanding and knowledge of the Kung Fu master taking advantage of a situation and turning it into something amazing. This is the case with many unconventional weapons found in the Chinese martial arts, especially those that are from the southern provinces with layman practioners. The Tiger Head Fighting Bench is such a weapon, or rather a tool, an everyday item found in the home or teahouse of the day. The bench is sturdy, hard and unforgiving when used properly in the master's hands. It crosses many different concepts of other weapons amalgamated into this tool that is now reborn for combat.

Our fighting bench is born of the movement and action of the tiger, full of power, strength and energy but balanced with agility, skill and intelligence. This makes the Tiger Head Fighting Bench a unique weapon among the ancient weapons of Kung Fu.

About the Author

Master Paul Koh is internationally recognized as one of the world's leading Kung Fu teachers. He has dedicated his life to the study of this ancient Chinese art form, immersing himself in Chinese language and culture since his early teens. He is one of the only major advocates for preserving and promoting this cultural art which, despite being frequently overshadowed today by popular fads, is a timeless and ever-evolving art that goes far beyond a punch and kick.

With over 40 years of experience, Master Koh has extensively trained with world-renowned Kung Fu masters in 少林洪家虎鶴拳術 Hung Gar Tiger Crane and 少林黑虎門虎爪派 Fu Jow Pai Tiger Claw Kung Fu. He has expertise in classical Chinese weaponry and in traditional Southern lion dancing. Master Koh is not only a martial artist who has dedicated countless hours, days, months and years to the study of his discipline, he is also a true educator, mentor and motivator. He has taught thousands of students and currently maintains a martial art training hall in NYC Chinatown. His knowledge of Kung Fu goes far beyond the physical execution or application of a technique, and extends to the historical and cultural background of Chinese martial arts, as well as the deep philosophical bedrock upon which this art form depends. These qualities, combined with an ability to speak and write clearly and eloquently, make Master Koh a unique phenomenon in the martial arts world.

In addition to studying, researching and teaching the Chinese martial arts for nearly four decades, Master Koh is an accomplished author, having written several previous texts on the art of Kung Fu, as well as having published many articles in various martial art magazines throughout the years. He currently serves as martial arts consultant on a mystery novel series published by Soho Press and written by John Shen Yen Nee (former executive at DC Comics and Marvel Comics) and SJ Rozan (author of 18 novels, Edgar Award winner). The first novel, *The Murder of Mr. Ma*, was released in early 2024, and Master Koh's work was hailed by reviewers for his "incredibly orchestrated fight scenes" and "absolutely fabulous displays of martial arts."

In 2018, Master Koh launched Kung Fu In A Minute, a unique approach to documenting, promoting and preserving the traditional skills, attitudes, traditions and practice of the art of Kung Fu. Through Kung Fu In A Minute, he has published 14 titles with more forthcoming. The Kung Fu in a Minute library of videos, articles and manuals will further promote, preserve and protect this ancient art form.

Master Koh practicing the bench in China

Additional titles available from
Kung Fu In A Minute Publications and Master Koh

Dynamic Art of the Tiger Series

十獨手
Ten Essential Techniques of the Tiger

虎爪木人樁大師
Tiger Claw Wooden Master

八卦虎爪對拆
Tiger Claw Eight Diagram Fighting Set

虎爪八卦棍
Tiger Claw Eight Diagram Long Pole

黑虎爪單刀
Black Tiger Claw Single Saber

虎爪 虎尾三節棍
Black Tiger - Tiger Claw Three-Section Staff

黑虎大耙
Black Tiger Claw Tiger Fork

Southern Shaolin Series

嶺南拳術五行拳
Five Element Fist

嶺南拳術仙鶴拳
Immortal Crane Fist

嶺南拳術虎爪手法
Principles of the Tiger

猛虎鐵鎚
Fierce Tiger Iron Hammer

嶺南虎爪雙刀
Tiger Claw Double Knives

虎鶴對拆
Tiger & Crane Matching Set

E-Books

伏地虎尾脚
Tiger Tail Kick: Ground Fighting Technique

黑虎爪對拆
Black Tiger-Tiger Claw Matching Set, vol. 1

For more information, visit
KUNGFUINAMINUTE.COM

www.ingramcontent.com/pod-product-compliance
Lightning Source LLC
Chambersburg PA
CBHW041122300426

44113CB00002B/28